Time and the Multi-Universe
A philosophy of time and time travel

E. Hughes

Copyright © 2022 E. Hughes

All rights reserved. No part of this book may be reproduced or transmitted in any form or by any means, electronic or mechanical, including photocopying, recording or by any information storage and retrieval system without permission in writing from the publisher.

Love-LovePublishing—Madison, WI
Paperback ISBN: 978-1-7377052-5-3
Hardcover ISBN: 978-1-961823-09-9
eBook ISBN: 978-1-737705-2-9-1
Library of Congress Control Number: 2022909508
Time and the Multi-Universe: A philosophy of time and time travel
Digital distribution | 2022
Paperback | 2022
Hardcover | 2023

Time and the Multi-Universe: A philosophy of time and time travel is not only about humanity's philosophical relationship with time, but rather a fundamental understanding of how time functions in our society as a mental construct and as phenomenon occurring in our universe.

To time infinitum.

Table of Contents

Chapter 1: *What is time?* 1
Chapter 2: *When we die, the universe dies with us* 5
Chapter 3: *Black holes and the expanding universe* 12
Chapter 4: *Manipulating time* 18
A 13 month calendar 30
Bibliography 50

Chapter 1

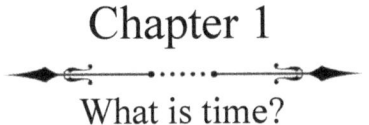

What is time?

What is time? Time is not merely a succession of events in terms of past, present, or future. In lay terms, and for the purpose of this book, it is a measurement. Time is the byproduct of a universe that is always in motion. Before time, before the Big Bang, there was nothing until something happened, setting the Universe—and time, in motion. Much of these events are theoretically already known to science. This book is a brief philosophy of time and the construct of time in our universe and daily lives.

<u>So what exactly *is* time and does it really exist?</u>

Short answer—yes, time does exist. Many believe that time is a construct (it can be) or an illusion (it isn't.) The notion that time does not exist is as unsettling as the notion that numbers don't exist. Numbers are immaterial but can represent material or immaterial phenomena occurring in the universe, whereas time certainly has both in its connection to gravity and space-time. While time and numbers are not macroscopic, both abstractions have a firm (or perhaps a metaphysical place in the universe. It's not so much what numbers are as to what they represent. Time and numbers are similar because they represent knowledge…information about something, somewhere, in the universe. Both represent a harmonious balance of metaphysical and material concepts with numbers as a mental construct and time as an observable and naturally occurring phenomenon of space, gravity, and motion. The two concepts must also work together because we need numbers to calculate time. They are very much part of the same family.

If life did not exist in the universe, would numbers exist? Would time exist?

"If an apple falls from a tree in a forest and no one is around to hear it, does it make a sound?" (George Berkeley, 1600s).

(Answer: No. The apple's impact on the ground produces vibrating particles [sound waves] but doesn't make a sound until the waves are interpreted by the brain.)

While numbers are metaphysically floating around in our minds to help us make sense of the world around us, the existence of numbers are also dependent upon our mental constructs to perceive them. For time to exist, time must also be observed or perceived by something or someone in the universe. Not only must time have an observer, but its duration must be experienced in some capacity. Imagine if the observer and the universe were both frozen in place... motionless, the natural environment unchanging. Erosion, decay, and the process of "aging" would not exist. Time would stop and the duration of "time" passing by would not be experienced by the observer because time does not exist in the universe without motion or gravity. The state of the universe would remain the same, like a photograph. Consequently, the observer would not exist at all in a universe without motion. Motion and heat in the universe causes all things to exist.

Imagine if the universe was in motion but the observer was *not* in motion...acts of nature caused by motions (terrestrial or otherwise) in the universe would act upon the observer. Over time, the duration of the observer's lifespan would undergo various stages of change and eventually come to an end. If, instead of an observer there was a rock or a mountain, the effects of time and change would eventually lead to erosion as the forces of nature is exerted on the rock or mountain. Time is duration and measurable change.

Where time and numbers differ is that numbers help us to observe consistent and reliable patterns demonstrated by material and metaphysical entities in the universe, while time can represent a dimension (Einstein), change, or the *duration* of a pattern or cycle in the universe. For Einstein, time as a 4th dimension is an intersection of where and when space and time meet. He wrote of gravity's effect on time, such as time dilation. The stronger the gravity, the slower time moves. Conversely, he wrote that time is relative to motion: the faster you move, the slower time moves. In a more relatable context, if I were to travel from California to New York by plane I would reach New York in a little over five hours. If I were to drive unimpeded from California to New York, it would take about 40 hours. The plane travels faster than the car, so time passes more slowly as I travel by air. Time is relative (Einstein). Earth's rotation (which is how we measure time on this planet) hasn't changed, of course. In this scenario time is moving slower in relationship to the velocity of the traveler. Time is the constant in this scenario while the velocity of the traveler flying, walking, or driving from California to New York is the variable.

Other scientists expounded on the nature of time along the lines of the Second Law of Thermodynamics and entropy, with the arrow of time pointing forward (Eddington, 1927). This means as a general rule that disorder increases with the passage of time and that systems will transition from order to disorder. I can agree with this notion. Time does have some constancy. Everything in the universe (apart from electromagnetic waves like light) unravels, erodes, decays, or inevitably comes to some sort of end or transformation. Stars explode, implode, or eventually burn out, sometimes taking the nearest planet with them. Does eternity exist? If we follow the arrow of time even the universe is subject to entropy. At some point in the history of the universe there was order until the Big Bang, which resulted in the disorder that gave us an expanding universe. The universe likely went from uniformity and order to disorder and will continue to sort itself until equilibrium is achieved.

Like stars and other matter in the universe, humans and organisms also die. Our bodies are subject to the arrow of time. From the moment we are born our bodies and even the life forms around us are constantly changing in cycles we can mostly predict by monitoring the sky. We know that the sky is changing, moving us in a consistent and forward direction (even if those changes and patterns appear cyclical), a cycle that takes us from night into day and from day to night. We know the sun will appear in the sky around a certain time each morning and if it doesn't appear, then something has come to an end—namely all of us. Where time becomes unpredictable, is that none of us can predict exactly *when* the universe will die. Or when *we* will die. Time can be vague, mysterious and unpredictable, but also inevitable.

If everything, including the universe as we know it, will eventually come to an end, what is the point of anything in life? Do *we* have a purpose?

What was the spark that created life from a molecule? How did chains of molecules evolve from a nonliving system into a system that developed the ability to replicate, into something primordial, into something that developed an innate instinct to survive, evolving over time until it developed mobility, the ability to consume, communicate, and even acquired a consciousness? 17th century philosopher Thomas Hobbes would say that human beings are matter and that metaphysical part of us (like thoughts, choice, or free will) doesn't exist and that our actions are governed by chemicals in our bodies reacting with a very physical brain. To Hobbes, our lives, thoughts and personalities are merely atoms, reacting.

As human beings reproduced, evolved, and spread across the planet, our numbers and actions became more destructive and chaotic towards each other and the planet. With the invention of nuclear weapons humans are capable of destroying life on Earth. As we become more disordered and chaotic we alter the planet in ways that will lead to the cycle of life ending for most of the species that live here. On Earth, the arrow of time marches us forward from order to disorder and chaos.

To demonstrate the Second Law of Thermodynamics, a scientist might put gas into a glass bottle and heat it until the gas was under so much pressure that it would erupt into disorder (at the molecular level), compelling motion. This is much like the buildup of carbon in our atmosphere where carbon gases continue to build until it traps heat coming in from the sun, creating a greenhouse effect. This is a direct cause of climate change as a result of activities by humankind. Are people evidence of disorder occurring on a planet that evolved to support life? If systems on Earth evolved to sustain life, is the evolution of human beings, the development of destructive forms of technology, and the eventual build-up of carbon in our atmosphere the arrow of time at work?

The biggest misconception about time is that time is an illusion. While time relies on our awareness of it, it still exists without our perceptions. However, we need consciousness to perceive time and our reality. In philosophy, time, perception, and reality are sometimes believed to be an illusion. But time is very real, very much part of a physical event in space between celestial bodies. *Time* describes what is taking place as well as when it is taking place and how these events intersect with the planet and ultimately, our lives. I don't have to see time at the macroscopic level to know it exists. I feel it as I age and my physical body undergoes various degrees of change. My body isn't time anymore than the sun, Earth, or space itself is time. Time is the measurement of phenomena.

Time is happening all around us, all over the universe, in the same moment, relative to events taking place in the cosmos. We share the same sun in our solar system, but time on Earth is different from "time" on other planets. Each planet is trapped in an orbit caused by the same distortion of space caused by the mass of the sun, the gears of time shifting very differently among them with distance resulting in lengthier revolutions. For example, a minute, hour, or even a day is different on Jupiter. A year on Jupiter is 11 Earth years.

Chapter 2

When we die, the universe dies with us.
Time Travel

We know that time does not move backwards. Time is directional—always flowing forward (Arrow of Time, Eddington, A).

Thus, time travel into the past is physically impossible because the past is a fixed point. The past as a state of physical reality doesn't exist. The past is metaphysical, existing only through the awareness of conscious beings. The past is a mental recollection or record of events that took place in the present. Therefore, only two states of reality exists, and that is the present and the future. What "is" and what will "be." The past is *not* a fixed or objective *reality*. It is fluid and can be changed by our perceptions. The "past" cannot be visited in a physical sense, but can be altered in a realm I call our meta-space, which is in the metaphysical space that houses our thoughts, memories, and perceptions. We all collectively exist in this meta-space as we are bound by the same threads of reality. Aspects of our reality is shared and then there are separate aspects of our reality that is unique to our individual perceptions. Some animals and insects exist in a different metaphysical space and perhaps, cannot even see us but experience us as natural phenomenon. They are bound by a different thread of reality. Single-celled organisms and other microorganisms share a different universe and reality. We have many universes, a multi-universe that we all live in together, but also separately. We can't read each other's thoughts or see through each other's eyes, but we have a collective and rational understanding about what is in nature and the universe. However, there are layers to reality that exist outside of our shared experience. This includes animals, insects, planets and microscopic organisms that exist on a different plane of reality in a part of this multi-universe that human beings will never experience. Imagine marine animals that have only lived in the sea. The sea is so vast and so deep, that there are mountains below the surface that are taller than the mountains we see above on dry land. The air above is as mysterious to most marine life as space is to people. For marine life, the sea is its universe.

To bacteria that might live in my intestines, I am the universe while the universe everyday people see and live in, might belong to an even larger universe that is so far beyond our comprehension that we will never have perception of it. We have an agreed upon reality but can never truly confirm what exactly *is* real, what exactly is "there" because it is all information interpreted by our individual brains in our individual meta-spaces. The universe, as it appears to me may look completely different to someone else. We can never confirm that what we "see" is the same. When we die, the universe dies with us. There is no real evidence that my perception of the universe exists beyond me or without me.

But time, no matter where we are or what our perceptions of reality are, has a purpose. Time binds us to a physical universe and reality. We can count on certain events to confirm within ourselves that the universe is not only in motion, but is a part of our reality.

While the past is a fixed point in our physical universe, in meta-space our thoughts of the past DO exist. The past and present *can* exist in this metaphysical space at the same place at the same time. Our physical bodies remain in the present (the larger universe) but our minds (in metaphysical space) can re-visit the past over and over again. Otherwise, what evidence do we have of the past if not our recollections or the collective and rational understanding that something came BEFORE now? Here in this meta-space, the past is an impression, seen from afar where "time" has even lapsed in our memories, growing weaker with time and distance. Physics in meta-space works similar to physics as it exists in the larger universe but abides by a *slightly* different set of rules and laws. In the metaphysical version of space, the past is captured by the light and the energy in our bodies, like a photograph, creating a record of events that we can revisit mentally. In a physical universe, there is no past. The past is merely a metaphysical recollection of an event that took place in the present.

Light from stars that shone a billion years ago appear to us in the night sky...in the present, much like the sunlight that left the sun a full eight minutes ago, before we see it here on Earth. We tend to discuss time travel in metaphysical constructs rather than physical constructs despite the fact that time is the byproduct of a physical event. To physically go back in time, we would not be able to reconstruct matter, gravity, energy and space-time as it was when past events originally took place. This is why we think of time travel in a metaphysical sense. Too many moving parts must be unraveled and rearranged across the universe to think of time travel to the past in a physical context. So how do you separate time travel from the physical events that give us time?

Time Reversal is a theoretical event in which time would reverse or run in a backwards motion. This differs from entropy as time would not march forward, but in the other direction. Time Reversal is achievable in mediums such as film or invariant studies in quantum physics involving subatomic particles. But in practice, time reversal is improbable for the purpose of time travel. It is impossible to restore or reverse matter back to its original state to a particular frame of space and time. And if it were possible to restore matter to its original state, would we also have the ability to restore intangible properties like our minds or consciousness to the state it was originally in at the point of the reversed timeframe?

The simplest mechanical definition of time, is that time is the measurement or duration of a massive body's influence on the objects around it, typically in space. In Albert Einstein's Theory of Relativity, massive celestial bodies can cause a distortion that warps the fabric of space, resulting in a force called "gravity". This warping can affect how objects within a certain distance move around these massive bodies. Objects like planets, moons, stars, and comets, become trapped by gravity and remain in a cosmic freefall as they orbit until they meet an inevitable end, which of course, would take billions of years.

The Theory of Relativity is an expansion of Einstein's previous theory, Special Relativity, which previously explained the connection between space and time but failed to explain gravity's role in this connection (Space.com). Einstein explained time as a fourth dimension. Space is where you are. Time is "when" you are. The two are entwined as you can't have one without the other, much the same as you can't have time without motion or gravity. Conversely, what is time without an observer? On our planet, time is a predictor of natural phenomena. Motion in our sky tells us when the sun rises and falls. Time is a predictor of the four seasons (two solstices and two equinoxes). Time tells animals when to hibernate, when to forage or produce offspring. It tells people when to plant or grow certain foods. It is time that tells trees when to bud. Plants, animals, and people need time to survive. Time regulates our lives and can even regulate our biology. Like the circadian and diurnal rhythms in people and animals. The circadian rhythm is an internal biological clock, and like the rest of the planet, can be affected by sunlight. Light tells your body when it's daytime (and that you should be awake) or when it's nighttime (the natural time for your body to sleep). The circadian rhythm follows a 24 hour cycle, roughly the same amount of time (duration) that it takes the earth to spin on its axis as it orbits the sun.

But when we think of time in our everyday lives, we think of time as an event, something that has taken or is taking place in the past, present or future. We look at our watches. We look at our phones. We look at the clock on the wall. If time tells us when we are and space tells us where we are...where does the past meet with time this in intersection? *Is time travel possible at all?*

The time travel paradox is that the past and present cannot simultaneously exist in the same moment. Whenever you are, you are always physically in the present. You also cannot simultaneously exist in the future and present at the same time. Once you are physically in the future, the future becomes the present.

It is possible however, to time travel under certain conditions. We may not think of it this way but we manage to time travel every day. The past is a fixed point, so travel to the past will always remain impossible, except where we store the past in our minds as memories. But there are other ways to visit the past, which is to capture it while still in the present. One day, I was watching an old movie from the 1930s, *Rasputin and the Princess*, and became curious about one of the actors. The actor was a man who looked to be in his 60s. I found him on a Hollywood database and learned the man was born in 1874. Another actor in the movie, Margaret Mann, was born in 1868. Another actor in the movie, Gustav von Seyffertitz, was born in 1863. As silly as it sounds, I was amazed that in the 21st century (*in the year 2021*), I was able to peek through time via recorded video (rapidly flashing still images) and see someone who was not only born (in 1863) but was alive in the second half of the 19th century into the 20th century. This person was born 158 years ago, when technology was still not as advanced as it is today. Maybe it's just me, but it's mind-blowing when put into perspective. But the movie was only a fictional snippet of the past rather than real-life events playing out on film. Imagine what we could do now if we had the ability to record every moment that takes place on Earth. Imagine what that information will mean to people in the future. A film from the 1930s captured a moment in time that I can revisit over and over again. *But what about time travel to the future?*

Time travel to the future is possible, but in a different way. Imagine I'm in California and I need to communicate with someone in New York, over 3,000 miles away. If I were to drive, it would take between 40 to 50 hours. By flight, a little over five hours. On foot, quite a long time. However, I can arrive in mere seconds by phone, by transmitting the sound of my voice via analog across the country using radio waves. I can also travel by radio waves into compressed and decompressed analog in the form of video communication via a cell phone or computer. Like the movie from the 1930s, video communication today uses still images that flash at a specific frame rate allowing the viewer to see images of the other party in what appears to be real time. I can metaphysically transport myself via video across the country in seconds. Not only can you hear my voice, but data transmitting my image (digitally or via analog) will be conveyed onto a LED screen that projects photons of light allowing the person on the other end of the video call to see me, via more than one form of energy. This is a form of time travel.

While I cannot transport my physical body (matter) across the country in seconds, I can slice through time, transport images and the sound of my voice, and project an image of my metaphysical self to the other party. I am also interacting physically, as sound waves from my voice interact with the other party's very real ear drum. This is a form of time travel as there is the effect of my being in New York and California at the same time. I am able to physically exist in California but also metaphysically (visually) appear in New York without physically being in New York. Instead of 40 hours of travel to get there, radio waves that travel at the speed of light has cut this time down to seconds.

In consideration of "physical" time travel, one must become mass-less. This would allow us to move through space in the form of energy like visible light or radio waves. Imagine if we used radio waves to transmit vibrations that were so strong it allowed us to physically touch across time and space. This may come as a surprise to some but radio waves have been used to transmit video calls since the 1930s. These radio waves use towers, antennas, and even the ionosphere, a region above the Earth that bounces sound back to ground level to carry sound waves around the world. The cosmos also send radio waves to earth from other planets, the sun, stars, and galaxies. Radio waves seem to be the most efficient and logical way to travel through time. Like light, radio waves are a form of electromagnetic radiation that is used to carry sound waves to a phone, radio, or broadcast, after which, the radio wave is discarded as the sound waves are transmitted through analog into the receiver's end of the phone. Sound waves, via radio waves allow you to hear, while regular light is a form of electromagnetic radiation that we can see. Radio waves are not to be confused with sound waves, though they work together with modern technology.

Scientists also use radio wave telescopes to detect and monitor black holes, quasars, and other radioactive bodies in the universe. Scientists can also convert radio waves from space into images, allowing them to look at black holes... which brings me to my next topic. Are black holes responsible for the creation of the universe? If so, how?

Chapter 3

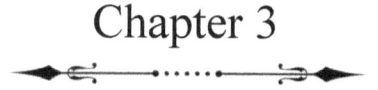

Black holes and the expanding universe

Einstein told us the mass of large bodies can warp the fabric of space, creating a curvature he called space-time, which affects how smaller bodies move around massive objects. When smaller bodies become trapped in an orbit around these massive bodies, it's called gravity. Gravity has a range that grows weaker with distance and accelerates or strengthens when an object is nearest to the source of gravity. Gravity is also held together by magnetic fields. The movement of those bodies is how we calculate or measure time.

Why do objects "orbit" around massive bodies instead of crashing directly into them?

Imagine sitting on a sofa or a bed and the cushion sinking under your weight. The cushion sinks in, warping the fabric around you, creating a dent. But the cushion is only warped within a certain radius of your mass/weight. Now imagine sitting a round object near the sunken in part of the cushion close to where you are seated. The object will roll directly towards you unhindered. In fact, it will crash into you and stop. If you sit the object just beyond the radius it will not roll towards you unless you disturb it by moving or shifting. Even further away, the object will not be affected at all. Now imagine if you were spinning or rotating very quickly. This would cause the object within your radius to fall towards you, but the descent would be delayed by your body's spinning motion, which is twisting the fabric in a rotating motion, slowing down the object's freefall towards you, drawing the object into an orbit around your body.

Figure 1 – space-time and curvature of space

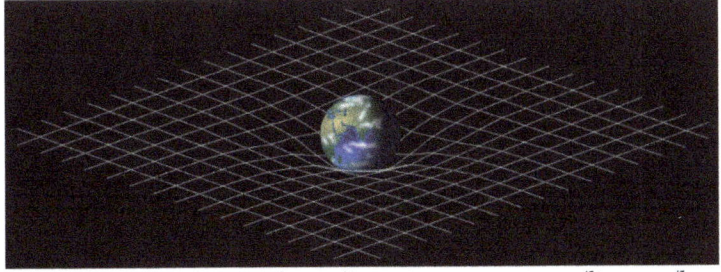

O Mysid, CC BY-SA 3.0 *https://creativecommons.org/licenses/by-sa/3.0, via Wikimedia Commons*

Take our own planet for example. Earth spins on its axis once a day, which is how we calculate a full day on Earth (more on this later). The sun also rotates and spins on its axis. The sun is in motion and moving through space. As the mass of the sun warps the fabric around it, it trapped the planets in its gravitational pull, drawing the planets towards it in a freefall. The planets should crash into the sun but the sun's spinning motion rotates the fabric below its mass, moving everything in the solar system in a counter-clockwise orbit around it, preventing Earth and closer planets like Mercury and Venus from crashing directly into the sun. Imagine the spin of a ball around a roulette table. The ball rotates against the wall of the table until the ball begins to slow down. The ball eventually loses momentum, succumbs to gravity, and rolls forward into one of the slots. Earth, along with the other planets would roll directly into the sun should their orbits around the sun lose momentum.

When large celestial bodies at the center of a galaxy rotates, it spins everything in the galaxy around it in a disc-shaped orbit. The same occurs around a black hole. Black holes are so immense, so incomprehensively massive, that it not only curves the space around it, but pulls anything within its radius (the event horizon) into it, almost like unplugging a drain and watching everything swirl into it. Black holes draw orbits around it in a disc-shaped spiral like other massive bodies.

Why are black holes black? A common answer among scientists is that the gravitational pull is so great that not even light can escape. Naturally, I agree. Astrophysicists Stephen Hawking and Albert Einstein have several widely accepted and widely respected theories on black holes.

Imagine a body so massive that it can't support its own weight. The body collapses inward like a giant sinkhole. I imagine it's like unplugging the drain in a sink, and everything around it, including dark matter (the dark stuff in space holding the universe together) getting sucked into the black hole. Matter, gas, and even light also begins to swirl around the black hole, eventually draining into it.

Imagine that this body is so utterly massive that it can suck entire stars, planets or even an entire galaxy into it. Instead of landing on top of this body (which is now a massive sinkhole), or orbiting around it, you are sucked into a void, or perhaps the internal temperature is so high that anything sucked in (including light) is superheated by the immense pressure and destroyed. Of course, this would defy the First Law of Thermodynamics which is that energy cannot be created or destroyed, it can only be transferred from one form to another. Famed astrophysicist Stephen Hawking reconciled this with Hawking Radiation. But in my thought experiment I pictured it in a different way.

Figure 2. Black hole, artist interpretation

By ESA/Hubble, CC BY 4.0,
https://commons.wikimedia.org/w/index.php?curid=49042636

Imagine there's a supermassive star (a star thousands of times the size of our sun). Imagine if this supermassive star is rotating on its axis. Suddenly that rotation begins to accelerate, spinning so fast that it nears the speed of light. This hyper-fast spinning motion not only warps space, but spins so fast that it flings the star's plasma, ionized gases, metals, and light into a disc-shaped orbit around its now-darkened possibly magnetized metal core. Not even light can escape the intense gravitational pull of the star's spinning motion. From space, it appears to us as a black hole because of gravitational lensing (an optical illusion where it appears that light wraps around a massive object in space), when it's actually a supermassive star shedding most of its hot exterior mass. This motion, along with the star's dense mass causes such a distortion of space, that it obliterates anything in its path, tearing these objects apart, collecting rocks and other debris, into its super-heated orbit.

Figure 3. Milky Way Galaxy

By Pablo Carlos Budassi - Own work, CC BY-SA 4.0,
https://commons.wikimedia.org/w/index.php?curid=102333184
Caption: Map of the Milky Way Galaxy with the constellations that cross the galactic plane in each direction and the known prominent components annotated including main arms, spurs, bar, nucleus/bulge, notable nebulae and globular clusters.

The spaghettification effect of black holes fades with distance. Rocks cool and new stars and planets form from left over remnants of this supermassive star. Over time, billions of years later as debris at the outer-edge of the supermassive star's spinning disc cools, rocks and dust collected by the super-fast spinning motion of the star has created a new galaxy, with new stars, planets, moons, comets, and asteroid fields. This could explain why there's a black hole at the center of every galaxy. The closer objects are to the black hole, the faster those objects spin around it. Over time, objects on the outer edge of the galaxy's arms move further and further away from the depleted star, expanding the universe. There are trillions of these expanding galaxies in space. With the black hole spinning at a superfast rate, time moves much differently near these black holes.

Time as a measurement is the speed and duration of a body's orbit in space relative to the gravitational force exerted upon it, whether that body is the sun, a black hole, a planet or the creation of a new galaxy.

We set our clocks to these enormous cosmological events. In a sense, time feels like an illusion, like it is beyond us, because the scale of the universe is incomprehensible to the human mind. But every time we look at a watch, or meet someone at a certain "time" it is because of the events taking place in the dark ether beyond our skies.

Chapter 3

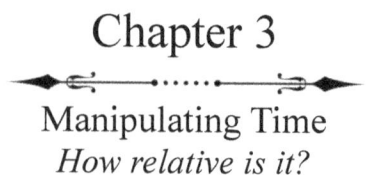

Manipulating Time
How relative is it?

When we think about time in the western world, we think about twenty-four hours in a day, seven days a week, 28, 30, or 31 days in a month, twelve months in a year, and 365.25 days and a leap year every four years.

This is not entirely correct. We don't have 24 hours in a day. A week is in fact seven days, but a month can only have 28 days...a maximum of four full seven day cycles in a month. No more, no less. A 31 day month is roughly just over a week and a quarter week long. To reach five full seven day cycles in a month, we would need a month with 35 days. So why do we have months with 30 to 31 days? We'll get to that next.

A month, depending on where you are, follows the lunar cycle. At some point in history, humankind was able to calculate time to better predict the seasons.

However, a month isn't 30 or 31 days and there are more than twelve months in a year. In fact, there are approximately 13 months in an Earth year. A secret month omitted from calendars as far back as 46 B.C. *How is this possible? There have always been twelve months.* The thirteenth month was always there. It was divided among the other twelve, giving them 30 or 31 days instead of 28. If you take any month with more than 28 days (months that have 30 or 31 days) and make them 28 days again, you'll have an extra 28 days left over as a result—an additional month. This means, every month of the year would have 28 days, and instead of 12months we would have 13 months in a year. A 13 month calendar is highly accurate and typically used to pinpoint the exact location at the same time, in our night's sky.

Figure 4.

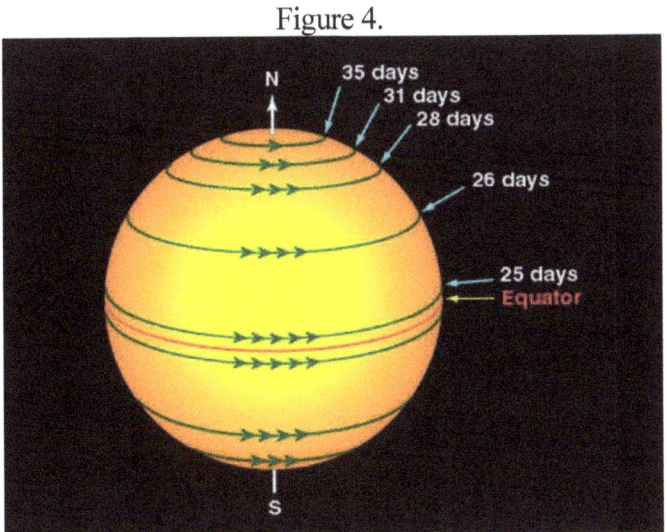

Image from https://www.nasa.gov/mission_pages/sunearth/science/solar-rotation.html

Why 28 days for a month? I'm not sure. I do know that the sun spins on its axis in one of the northern regions in 28 days. In an even higher part of its northern region, it rotates on in its axis in 31 days, and in the uppermost region it rotates in 35 days. The sun is not a solid mass like a planet, it's primarily made up of gas and liquid so different parts of its body rotate at a different speed. It would seem the 28 days might have been applied from the sun's rotation. This is called sidereal time.

The 13 month calendar is so accurate that it is commonly used by accountants in the United Sates. Using a 13 month calendar instead of the much flawed 12 month calendar would mean a year would be approximately 364 days instead of 365.25 days, with no leap years every four years. It also means that every day of the year would fall on the same day of the week every single year. Unfortunately, if your birthday fell on a Tuesday, it will always fall on a Tuesday. If we use a 13 month calendar the year is not 2022, the year is actually 2027.

So how do we calculate a 13 month calendar?

If there are 13 months instead of 12 then why are we using a 12 month calendar? The 13 month calendar has not been used in a very long time and was eventually forgotten in millennia long past. I think it's important to understand time to understand our place in the Universe. The best way to explain it to you would be to show you.

The Gregorian calendar we currently use now is the one of the most commonly used calendars in the world. There are 12 months in the Gregorian calendar as listed below in order:

January	February	March	April
May	June	July	August
September	October	November	December

However, January wasn't always the first month in a calendar year. One of the earliest calendars used in the western world was the Julian calendar. The Julian calendar was organized by Julius Caesar, but wasn't implemented until after his death in 45 B.C. Under the Julian calendar, March was the first month. April was the second month of the Julian calendar. May was third, June was fourth,
July was fifth , August was the sixth month (formerly

known under its Greek name; Sextiliss), September was the seventh month, October the eighth month of the year, November the ninth month, and December was the tenth month of the year.

The "Sep" in September refers to the number 7 in Latin, the "Oct" in October refers to the number 8, the "Nov" in November refers to the number 9, and the "DEC" in December corresponds to the prefix of the number 10 in Latin.

So how did the latter months come to be known as the 9th, 10th, 11th, and 12th month, instead of the 7th, 8th, 9th, and 10th, respectively? On the Julian calendar, the months January and February were the last two months of the year and not the first and the second, unlike the Gregorian calendar we use today. A Gregorian year is approximately 365.25 days, with a leap year every 4 years, excluding years divisible by 100. Leap years are divisible by 400.

The Julian calendar was a commonly used calendar until Pope Gregory XIII approved an updated version of the Julian Calendar in 1584. This updated calendar is referred to as the Gregorian calendar. You may wonder how this all ties into having a 13th month, or a "missing" month in a standard year. And where did that .25 days come from that resulted in a leap year every four years?

Let's first define what a "month" means. The definition of a month is the period in which the moon leaves one point to orbit around the Earth in a gravitational cycle until it arrives at the original point again. Generally a lunar cycle (or month) averages 27.9 days. A Synodic month has 29 days. A lunar cycle can vary marginally, depending on your location. I prefer to use sidereal time.

When you think of a month, think of 28 days. No more, no less. Not 30 or 31 days.

To begin, here's a bit of Q. and A. (Trust me, this is going somewhere)

Q. How many days are in a week?
A.) There are 7 days in a week.

Q. How many [full 7 day] weeks are in a month?
A.) There are 4 weeks in a month (no more, no less).

Q. 7 days times 4 weeks amounts to how many days in a standard month?
A.) 28 days

If there are approximately seven days in a week and only 4 *full* weeks in a month (no more, no less), then that means a month is only 28 days. The Gregorian calendar shows a total of 12 months. However, there are 7 months of the year with 31 days. There are 4 months out of the year with 30 days, and 1 month of the year (February) with 28 days excepting leap years. How do we reconcile a 13 month calendar to 364 days in a year?

Q.) If we change all 12 months of the year to 28 days, what happens to the days left over from months that gave us 30 or 31 days? If you add the left over days together from months with 30 or 31 days, approximately 28 days are left...a full month like the other 12, resulting in 13 months.

I'm a visual learner, so I'm going to use old long hand math to add things up.

Remember when I told you that there were 7 months in the Gregorian Calendar that uses 31 days instead of 28? 4 months that use 30? And 1 that uses 28?

IF 31 [days] times 7 [months] = 217 days.

30 days times 4 [months] = 120 days.

28 days times 1 [month] = 28 days

When we add all of the days together, we'll have a sum of 365 days!

Now if we divide 365 [days] into 28 [days] we end up with 13.0357214 months in a year (I'll explain how we lose that .03 margin a bit later).

Not only would we have an equal number of months, but each day of the week will fall on the same day each year, as well as the holidays. The question is why did the Julian and later the Gregorian Calendar abandon the 28 day month?

It's possible that a thirteen month calendar existed before the change by Julius Ceasar in 46 B.C. For example, the Mayans used a 260 day calendar. If you do the math this equates to just under 10, 28 day months. The Mayan calendar was based on a human gestational period (the period of a woman's pregnancy). In early Europe, superstition may have played a role in why 13 month calendars were avoided. In fact, Europeans even avoided the 13th astrological sign, Ophiuchus, by removing it altogether. Ophiuchus is the 13th Zodiac sign existing between Sagittarius and Capricorn. Astrological zodiac signs correspond to the constellations that occupy an area of the celestial belt surrounding Earth. Western astrological signs are an abstract area of the sky that follows the path of the sun over the course of a year. Ophiuchus would have correlated with the missing 13th month.

The astrological zodiac signs begin in March (formerly the first month of the year), with Aries. Because of the omitted 13th month and its correlating sign, Ophiuchus... Aries begins on the 21st of March and has a 31 day period just like the month it is supposed to represent. The other astrological signs in our calendar year are the same.

The sun signs represent all of the stars in Earth's celestial belt. These are constellations that include the sun at some point in its ecliptical field. There are 13 zodiac astrological signs: Aries, Taurus, Pisces, Leo, Virgo, Scorpius, Ophiuchus. Sagittarius, Capricorn, Aquarius, and Ophiuchus.

Ophiuchus is the only system not counted as a zodiac sign, even though it is one of the constellations that crosses the sun's ecliptic (the path the sun follows during the year). Ophiuchus is represented by a man with a serpent around his waist. He is thought to represent the Greek God of Health and Medicine: Asclepius. This image is also thought to represent the image of Adam in battle with the Serpent/Satan. Which could also shed some light on superstitions surrounding this Zodiac sign and why it is not in use. Whatever the reason, at some point in history an extra month was omitted by government and religious leaders and the world forgot.

Q. When I multiply 28 days times 13 months, there are only 364 days.

A.) There's a mathematical reason why. The reason a year is only 364 days under the 13 month calendar, is because a full day is based on the amount of time it takes Earth to rotate on its axis. However, contrary to what is commonly held belief, a full day is not 24 hours, but 23 hours, 56 minutes, and 4 seconds (23.934). If we calculate a year based on the exact amount of time it takes the Earth to rotate, we get an approximate answer of 364 days.

The reason we have 365 days under a 12 month calendar is because our clocks are designed to calculate 24 hour days and not its approximate number of 23.934 hours a day. This results in an additional 2 hours a month on our calendar. 2 [hours] times 12 [months]= 24 [hours]. An additional day a year, resulting in 365 days in a 12 month calendar.

To further illustrate this miscalculation I will first calculate a year based on the Earth's rotation at 23.934 hours in a day, followed by a calculation based on 24 hours. It looks harder than what's actually there. But if you read it, you'll get it right away.

It is likely that in 46 BC when Julius Ceasar made the Julian Calendar, and when Pope Gregory XIII approved the Gregorian Calendar, astrologists were unable to accurately estimate Earth's rotation at 23.934, and so, opted to round the number to 24, thus making it easier to make a calendar with 12 months, the mishap causing an additional day every four or so years.

On February 9, 1922, more than a hundred years ago, The United States of America's House Judiciary Committee held a hearing on "...the passage of time and how America kept track of it," (*History.house.gov*). This was called the "Liberty Calendar" and it divided the year into approximately 28 days and 13 months. Ultimately, the reform did not pass and was eventually forgotten.

The math:

There are 60 minutes in 1 hour.

There are a total of 1380 minutes in 23 hours.

1380 minutes = 23 hours

+ 56 minutes = 1436 minutes in a full day.

1436 x 28 days = 40208 [minutes] = a month

(4.091 [seconds] * 28= 114.548 seconds)

114.548 [seconds] - 60 [seconds] = 54.548 [seconds] =

add the additional 60 seconds as one minute to 40208 + 1 minute = 40209

M = 40209 and 54.548 seconds.

40209 / 28 days = 1436.055

1436.055 / 60 = 23.934.25 [HOURS] in a day

23.934 * 7 = 167.538 [hours] in a week. (OR simply 23.934 * 28 = 670.152 [HOURS])

167.538 * 4 = 670.152 [hours] in a month.

670.152 * 13 = 8711.976 [hours] in a year.

8711.976 / 364 = 23.934 hours = 1 day

Based on the accurate 23.934 hours in a day, there are only 364 days in a year.

ON A 24 HOUR DAY CALENDAR:

60 [= 1 minute] * 24 hours = 1440 MINUTES in a day

1440 * 31 = 44640 * 7 [months] = 312480 [minutes per year]

1440 * 30 = 43200 * 4 [months]= 172800 [minutes p.y]

1440 * 28 = 40320 * 1 [month] = 40320 [minutes p.y]
add MINUTES TOTAL IN A YEAR for each = 525600

525600 / 60 [MINUTES] = 8760 HOURS [in a year]

8760 MINUTES / 24 [hours] = 365 DAYS [A YEAR]

(8760 minutes divided by 24 hours calculates 365 days a year).

BECAUSE THERE IS AN EXCESS OF 2 HOURS per month due to rounding off to 24 hours from 23.94, we end up with a leap year.

2 [hours] * 12 [months] = 24 hours a whole additional day. So instead of an even 364 days a year that we end on the 13 month calendar, we end the year with 365 days on the 12 month calendar and .25 days as a result of the extra 2 hours a month.

Thirteen month calendar

Are we in the year 2022 or 2027?

Using the 13 month calendar are we actually in the year 2022? If there were 364 days in a year, on a 13 month calendar it is the year 2027.

2022 (years) times 365.242190 (days) = 738,519.70818 (days). Subtract 455 leap years. Divide 738,064.70818 (days) into 364 (days). The total is 2,027.6502971978 (years).

To close this philosophical journey on time, I am providing a 28 day, 13 month calendar at the end of the book. See if you can pinpoint your birthday.

MARCH

Sun	Mon	Tues	Wed	Thurs	Fri	Sat
1	2	3	4	5	6	7
8	9	10	11	12	13	14
15	16	17	18	19	20	21
22	23	24	25	26	27	28

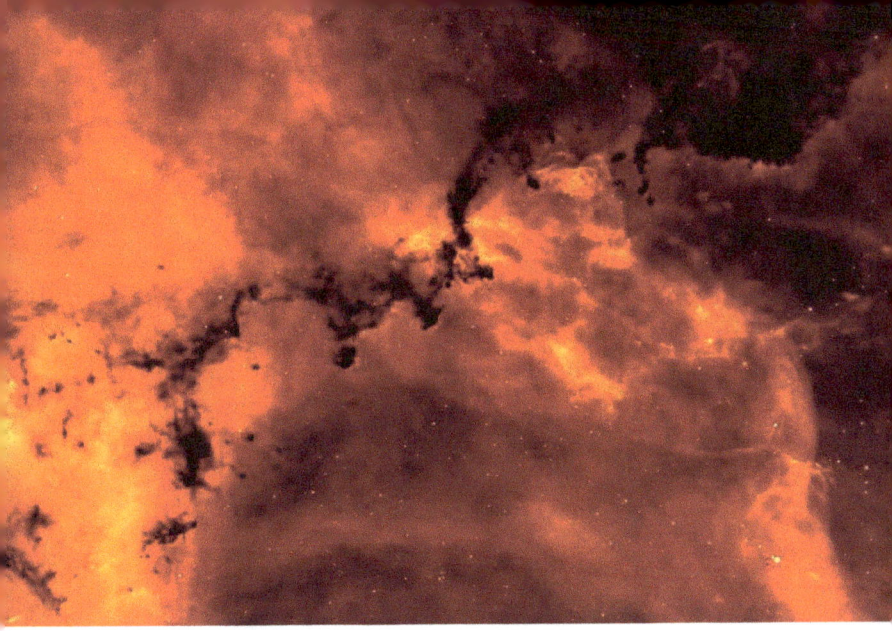

APRIL

Sun	Mon	Tues	Wed	Thurs	Fri	Sat
1	2	3	4	5	6	7
8	9	10	11	12	13	14
15	16	17	18	19	20	21
22	23	24	25	26	27	28

MAY

Sun	Mon	Tues	Wed	Thurs	Fri	Sat
1	2	3	4	5	6	7
8	9	10	11	12	13	14
15	16	17	18	19	20	21
22	23	24	25	26	27	28

JUNE

Sun	Mon	Tues	Wed	Thurs	Fri	Sat
1	2	3	4	5	6	7
8	9	10	11	12	13	14
15	16	17	18	19	20	21
22	23	24	25	26	27	28

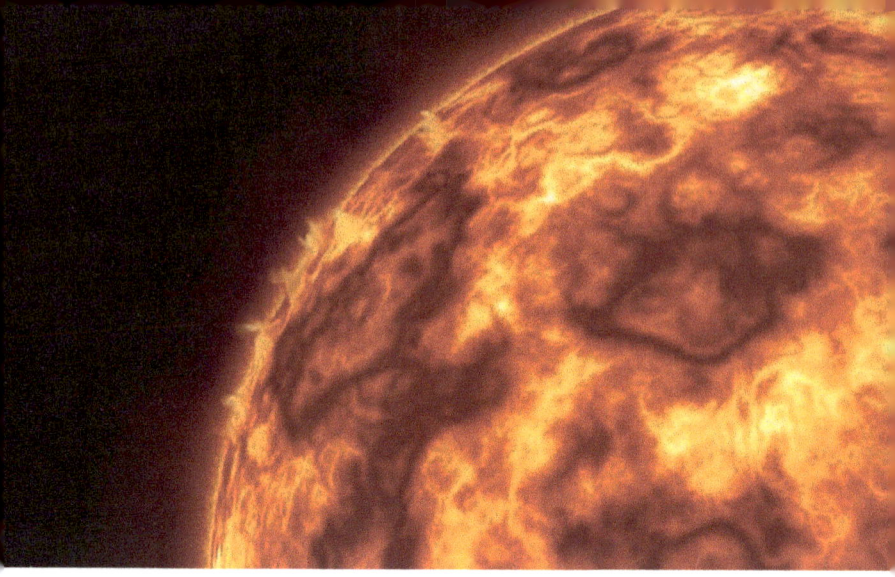

JULY

Sun	Mon	Tues	Wed	Thurs	Fri	Sat
1	2	3	4	5	6	7
8	9	10	11	12	13	14
15	16	17	18	19	20	21
22	23	24	25	26	27	28

AUGUST

Sun	Mon	Tues	Wed	Thurs	Fri	Sat
1	2	3	4	5	6	7
8	9	10	11	12	13	14
15	16	17	18	19	20	21
22	23	24	25	26	27	28

SEPTEMBER

Sun	Mon	Tues	Wed	Thurs	Fri	Sat
1	2	3	4	5	6	7
8	9	10	11	12	13	14
15	16	17	18	19	20	21
22	23	24	25	26	27	28

OCTOBER

Sun	Mon	Tues	Wed	Thurs	Fri	Sat
1	2	3	4	5	6	7
8	9	10	11	12	13	14
15	16	17	18	19	20	21
22	23	24	25	26	27	28

NOVEMBER

Sun	Mon	Tues	Wed	Thurs	Fri	Sat
1	2	3	4	5	6	7
8	9	10	11	12	13	14
15	16	17	18	19	20	21
22	23	24	25	26	27	28

UNDECEMBER

Sun	Mon	Tues	Wed	Thurs	Fri	Sat
1	2	3	4	5	6	7
8	9	10	11	12	13	14
15	16	17	18	19	20	21
22	23	24	25	26	27	28

DECEMBER

Sun	Mon	Tues	Wed	Thurs	Fri	Sat
1	2	3	4	5	6	7
8	9	10	11	12	13	14
15	16	17	18	19	20	21
22	23	24	25	26	27	28

JANUARY

Sun	Mon	Tues	Wed	Thurs	Fri	Sat
1	2	3	4	5	6	7
8	9	10	11	12	13	14
15	16	17	18	19	20	21
22	23	24	25	26	27	28

FEBRUARY

Sun	Mon	Tues	Wed	Thurs	Fri	Sat
1	2	3	4	5	6	7
8	9	10	11	12	13	14
15	16	17	18	19	20	21
22	23	24	25	26	27	28

Bibliography

Einstein, Albert (1919). "Fifteenth Edition, Relativity: The Special and the General Theory." Samaira Book Publishers; 1st edition (May 1, 2018).

Hignett, Katherine "Arrow of Time Reversed by Physicists in Quantum Experiments". Newsweek. (2017, December 17). https://www.newsweek.com/quantum-physics-reverse-arrow-time-correlation-729948

Jones, Andrew Zimmerman. (2021, March 10). "Does Time Really Exist?" Retrieved from https://www.thoughtco.com/does-time-really-exist-2699430

Lucas, Jim, Hamer, Ashley, (2022 Feb 07) "Second law of thermodynamics". https://www.britannica.com/science/second-law-of-thermodynamics

Lucas, Jim (2019, Feb 27) "What are radio waves?" https://www.livescience.com/50399-radio-waves.html

Kellermann, Kenneth I, (2021, Mar 29) "radio telescope". *Encyclopedia Britannica*, https://www.britannica.com/science/radio-telescope.

Kuhn, Robert Lawrence (2015, July 06) The Illusion of time: What's real? https://www.space.com/29859-the-illusion-of-time.html

O'Callaghan, Jonathan (2022, February 21), How It Works magazine." https://www.space.com/time-how-it-works

Tillman, Nola Taylor, Bartels Meghan , Scott Dutfield (2022, January 05) Einstein's Theory of General Relativity https://www.space.com/17661-theory-general-relativity.html

http://www.exactlywhatistime.com/physics-of-time/the-arrow-of-time/

https://history.house.gov/Blog/2020/February/2-28-Liberty-Calendar/

www.ingramcontent.com/pod-product-compliance
Lightning Source LLC
Chambersburg PA
CBHW041153110526
44590CB00027B/4215